I0063059

Mathematics Can Be Overwhelming

Copyright © 2025 Kyle Whiting

ISBN: 978-1-967360-00-0

First Edition: 2025

Written, illustrated, and published by Kyle Whiting
www.kylewhiting.com

This book is for informational and educational purposes only. While every effort has been made to ensure accuracy, the author assumes no responsibility for errors or omissions. Readers are encouraged to explore mathematical concepts at their own pace and discretion.

Dedicated to those ready to give up on mathematics. May this book provide you with a new curiosity.

Before you give up, please keep in mind:

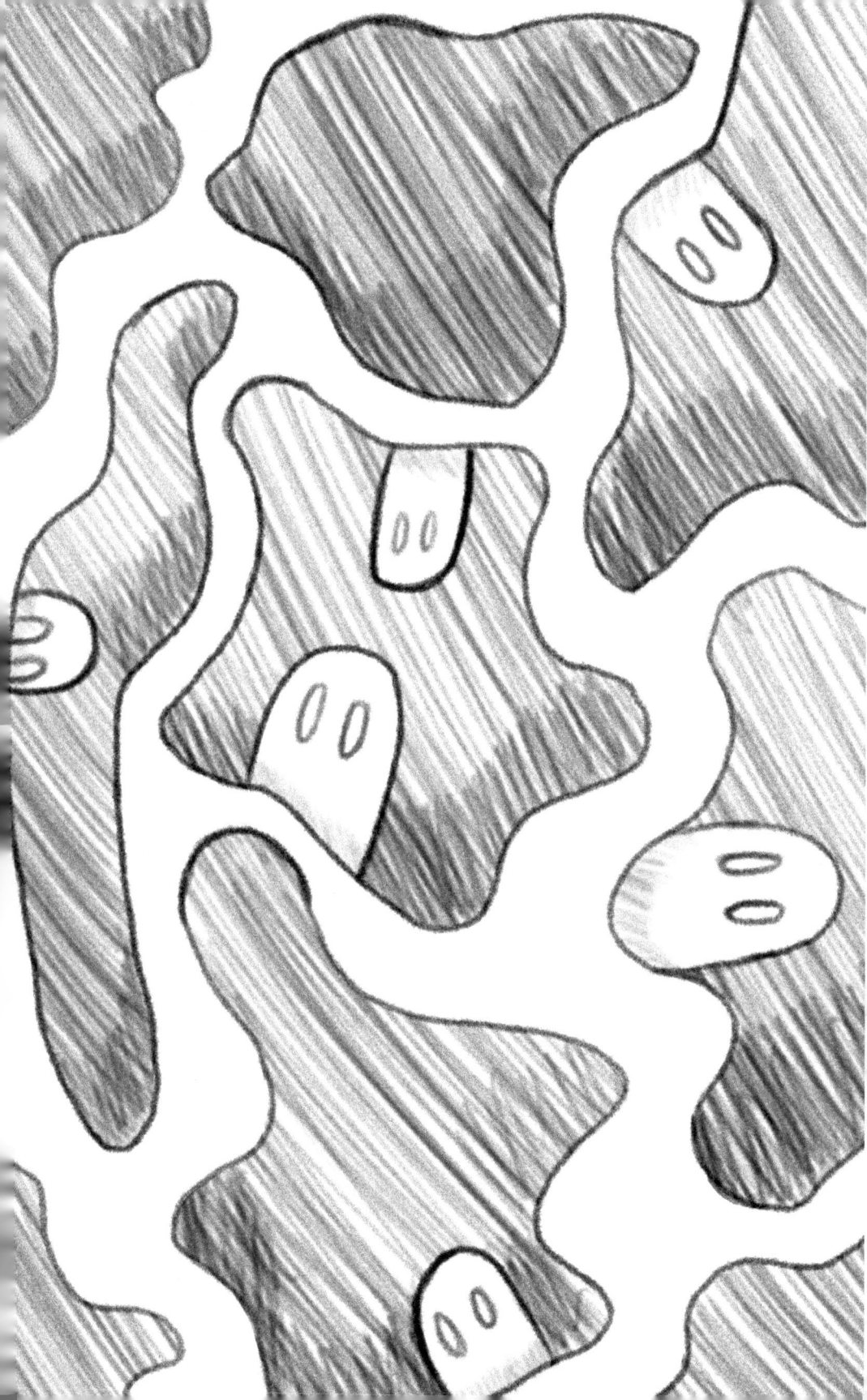

Mathematical progress moves at its own pace,
Yet in school, it's taught like a frantic race.
I believe there is value in taking a step back.
Allow me to illustrate and give you some slack.

It took billions of years for earth to form,
Millions more before humanity was born.
Time and evolution shaped two gifts—
To reason abstractly and convey what's missed.

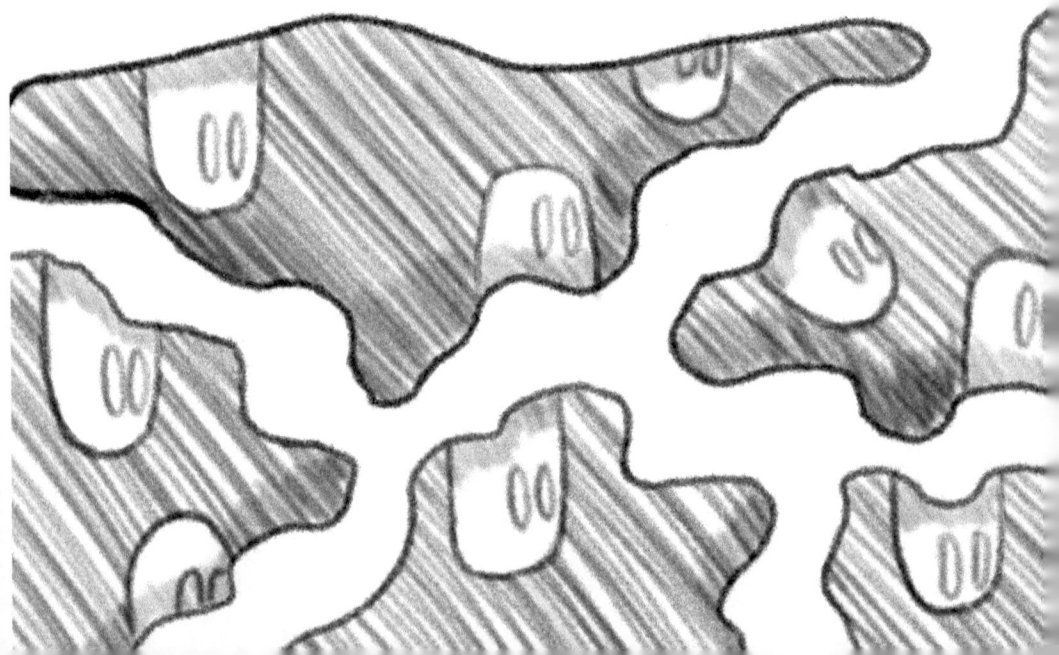

Reader, listen—can you see?
This gift was meant for you and me.

Could it be, this spark is in you?
You may not see it yet—but I do.

"Counting is simple," most declare in our day,
Yet for our ancestors, it wasn't this way.
Counting itself was a marvel to know.
In their time you would be the star of the show.

•   ••   •••   ••••   —   •   ••   •••   ••••   ═
‗•   ‗••   ‗•••   ‗••••   ══   ══   ══

I   II   III   IIII   ⊬⊬   ⊬⊬⊬   ⊬⊬⊬II   ⊬⊬⊬III   ⊬⊬⊬IIII

⊬⊬⊬IIIII

Ⲩ   Ⲩ   Ⲩ   Ⲩ⊻   Ⲩ⊻   Ⲩ⊻   Ⲩ⊻⊻   Ⲩ⊻⊻   Ⲩ⊻⊻⊻

I	II	III	IV	V	VI	VII	VIII		X
XI	XII	XIII	XIV	XV	XVI	XVII	XVIII	XIX	XX
XXI	XXII	XXIII	XXIV	XXV	XXVI	XX			

INK

1	2	3	4	5	6	7	8	9	10
11	12	13	14	15	16	17	18	19	20
21	22	23	24	25	26	27	28	29	30
31	32	33	34	35	36	37	38	39	40
41	42	43	44	45	46	47	48	49	50
51	52	53	54	55	56	57	58	59	60

In the counsel of arithmetic forged long ago,
Persians compiled a work many would know—
Answers to questions from far and wide,
By Babylonians, Egyptians, and others who tried.

Answers were found, but questions remained,
Allowing mathematicians to continue untamed.
If Persians of old could attend your class now,
Do you think they would wonder, How?

We sprint to the finish, eager to be done.
Arithmetic checked off. Algebra begun.
Don't fall prey to this false start—
Steady your mind, let reason chart.

Calculus

Finish

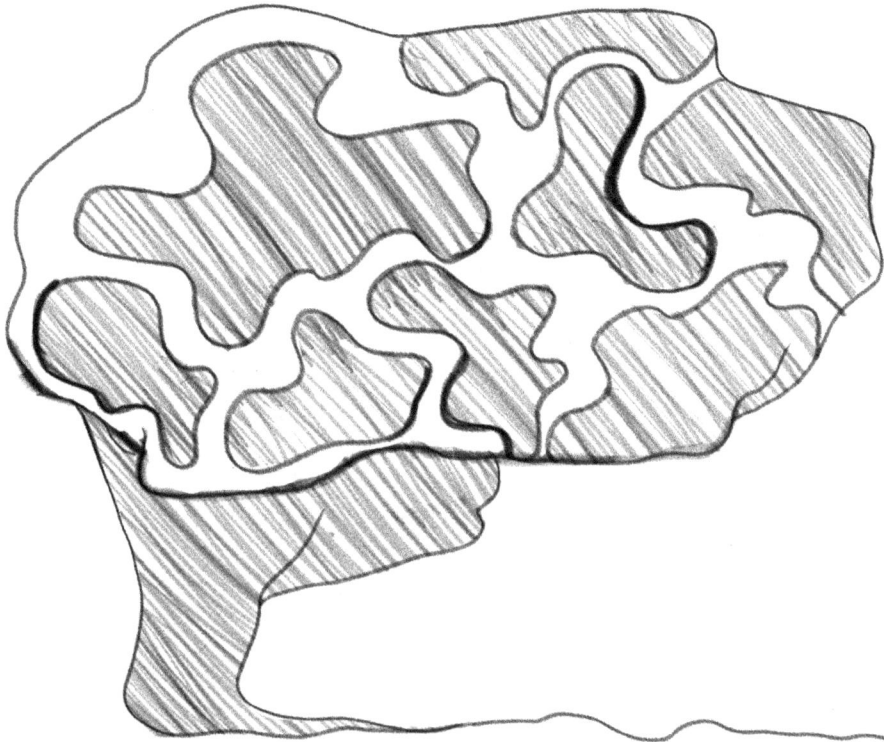

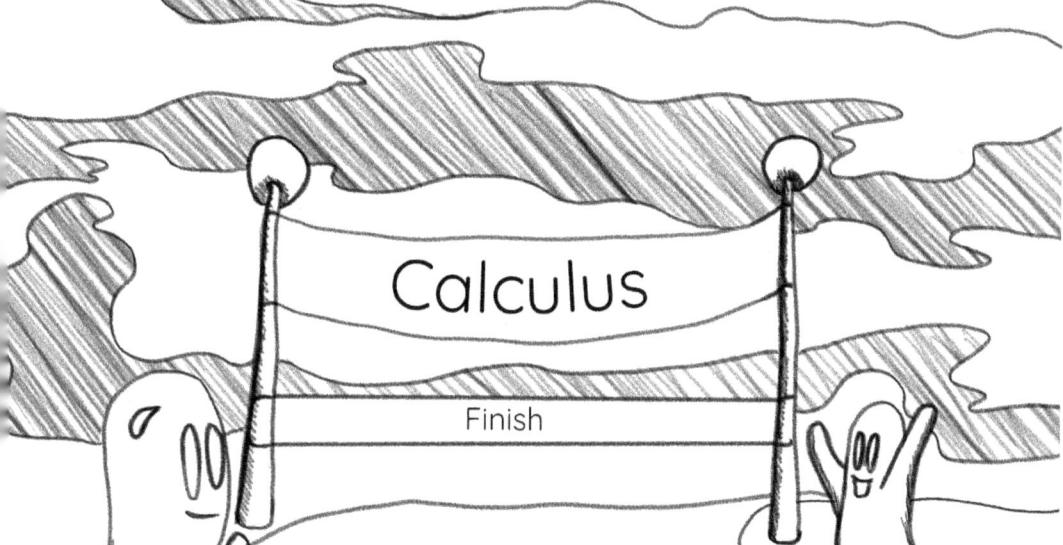

For if you're racing, you're
likely to lose
A precious gift that most
would choose.

Mathematics is not born from answers retold;
Its beating heart stems from questions untold.
One must acknowledge the subject would die
If questions were not allowed to thrive.

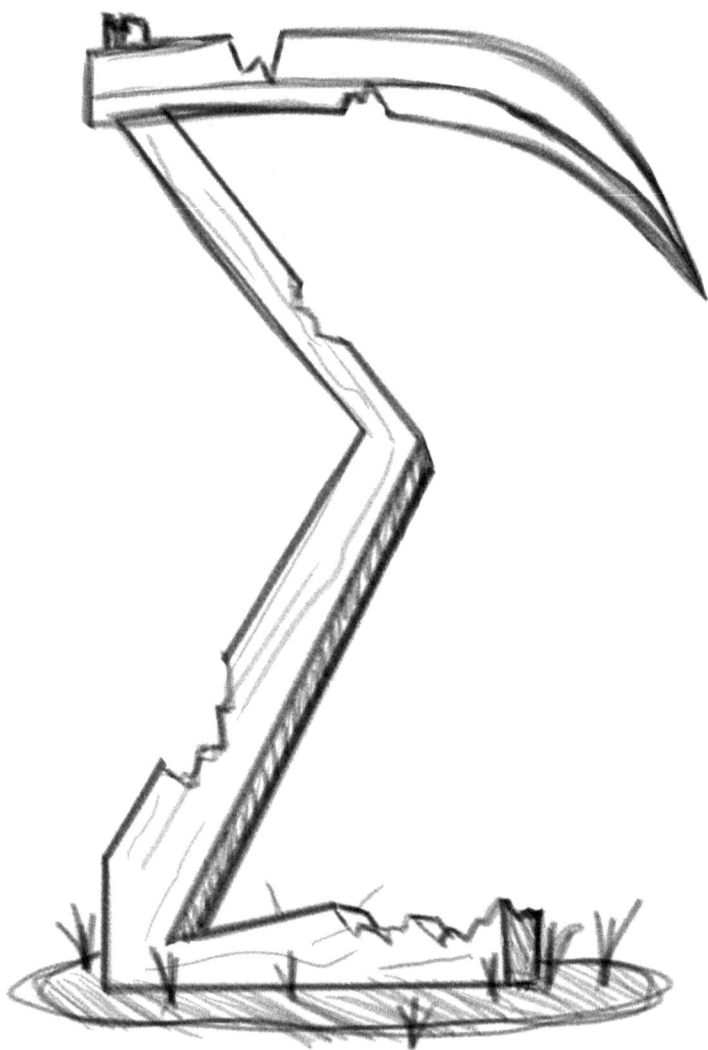

$x=-45$

$y=2$

45

$x(x)(y)=4$

87

3.14159

$x=2y+4$

Yet answers are given before questions appear—
No wonder so many find math unclear.
Do not fall victim to this misdirection,
Mathematics gives answers, but it's all about
questions.

Humans young and old ask questions with ease,
As if by nature we find reasons we please.
What sets math apart from legend and lore
Are the patterns and proofs that open new doors.

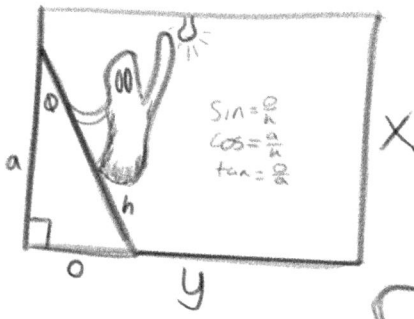

$$\sin = \frac{o}{h}$$
$$\cos = \frac{a}{h}$$
$$\tan = \frac{o}{a}$$

X

Y

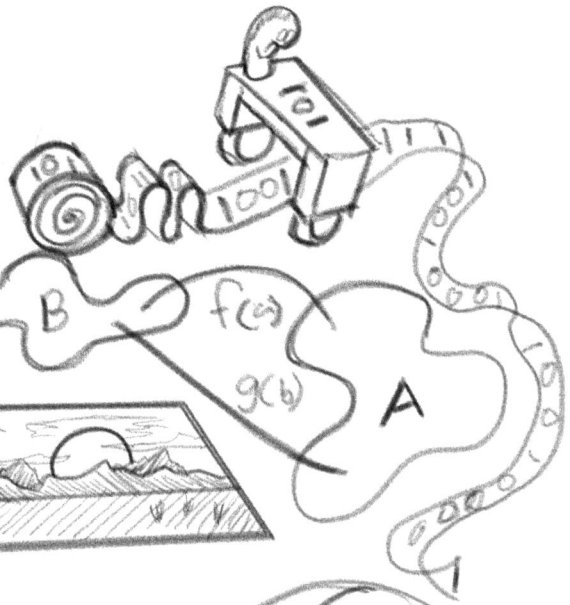

B

f(a)

g(b)

A

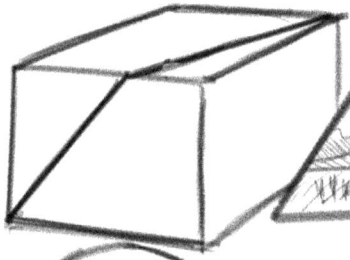

r

πr^2

$2\pi r$

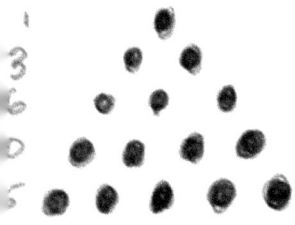

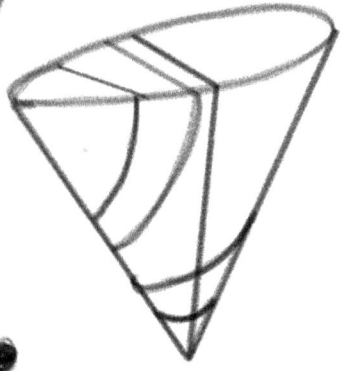

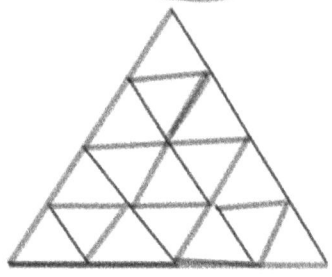

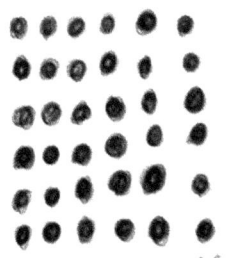

4 9 16 25 36

Your skeptical mind may still demand more,
So here's a question to explore:

How much of math, both old and grand,
Was shaped by human hands?

Katherine Johnson, who calculated paths to the stars

human.

Sir Isaac Newton, the father of calculus

human.

Al-Khwarizmi, the father of algebra

human.

Euclid, the master of geometry

human.

Hypatia, one of the earliest female mathematicians

human.

Babylonians, Mayans, Egyptians—

All human, just like you.

Dear reader, your humanity provides you with the tools to learn. I ask that you give yourself the time that our universities and schools may not.

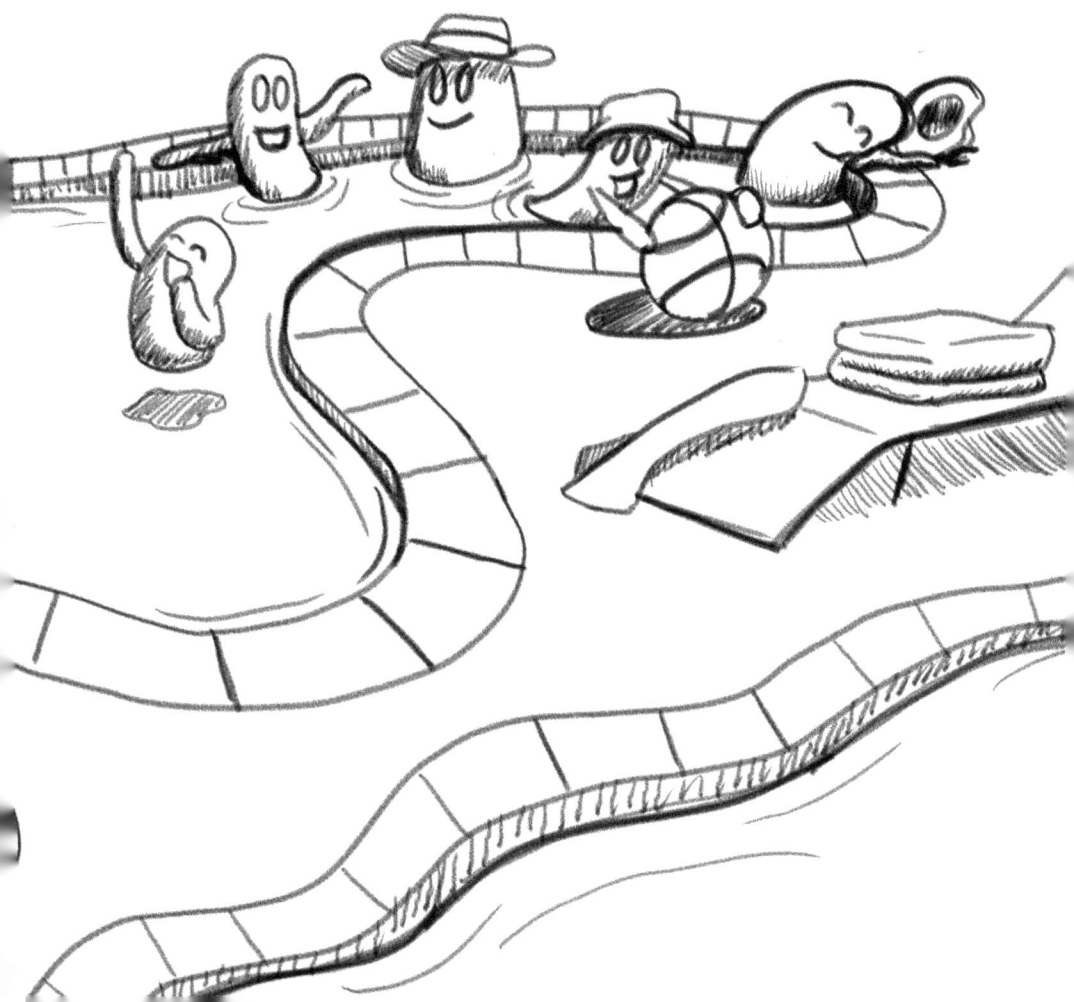

At the very least, give yourself as much time to learn a concept as the mathematician who discovered it.